CITES&Timber

T0136678

Ramin

L. Garrett, H.N. McGough, M. Groves and G. Clarke

2010

Ramin
Gonystylus spp.

Ramin (*Gonystylus* spp.) is a light tropical hardwood tree genus. There are some 30 species of Ramin, all of which are native to the peat swamp forests of Southeast Asia including Brunei Darussalam, Fiji, Indonesia (Kalimantan and Sumatra), Malaysia (Peninsular Malaysia, Sabah and Sarawak), Singapore, the Solomon Islands and the Republic of the Philippines.

It is in significant international trade which is virtually restricted to wood of *Gonystylus bancanus*, predominantly in parts and derivatives. Malaysia is the dominant exporter, with small volumes coming from Indonesia. Major importers include the European Union (dominated by Italy), the People's Republic of China and the USA. Re-exports of products are also important in the trade, in particular from the People's Republic of China, Italy and Singapore.

The global trade in Ramin is reported to be worth more than USD $100 million per year. The trade is in a wide range of material including sawn wood, semi-finished products (mouldings, dowels) and finished products. Finished products include furniture parts, flooring, curtain rods, umbrella poles, futons, snooker and pool cues, tool handles, technical drawing implements, window shutters, slatted wooden blinds, picture frames, slatted wooden doors, and veneers. The CITES listing regulates all of these products.

Scientific names

The full genus *Gonystylus* is regulated by CITES.

There are some 30 species in the genus. Of those species found in large scale commercial trade, *Gonystylus bancanus* (Miq.) Kurz. is the most heavily exploited.

Trade or common names

English	Ramin
French	Ramin
Spanish	Ramin
Indonesian	Gahara buaya (Sumatra, Kalimantan), Medang keladi (Kalimantan)
Malaysian	Gaharu buaya (Sarawak), Melawis, Ahmin, Kaya garu, Ramin telur
Philippines	Lanutan-bagyo, Anauan

Ramin in trade

Current trade in Ramin originates from two of its range States, Malaysia and Indonesia. This primary export is mainly in the form of timber and sawn wood. Any country exporting Ramin which is not a range State is a re-exporter. There are no plantations of Ramin outside the range States and all trade is wild in origin. Re-exports are predominantly finished and semi-finished products.

All trade data used in this guide has been sourced from www.unep-wcmc.org/citestrade/index.cfm. It should be noted that this data is subject to change. The data used in all examples covers the years 2005 – 2007. All trade is recorded in cubic metres. Trade recorded in units other than cubic metres is not included.

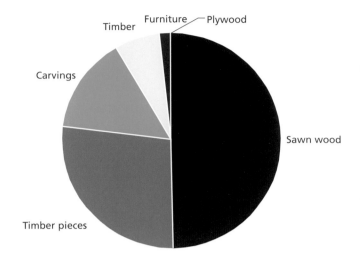

What is controlled under CITES?

The CITES Appendix II listing is annotated as # 1, in effect, this means all commonly traded parts and derivatives are regulated by the Convention and require CITES permits.

Gonystylus **spp. # 1** = designates all parts and derivatives, except:

(a) seeds spores and pollen (including pollinia);

(b) seedling or tissue culture obtained in vitro, in solid or liquid media, transported in sterile containers; and

(c) cut flowers of artificially propagated plants.

Note: This annotation may be changed at a forthcoming meeting of the CITES Conference of the Parties. However, this will not affect the regulation of all commonly traded parts and derivatives of *Gonystylus* spp.

Ramin parts and derivatives frequently seen in trade:

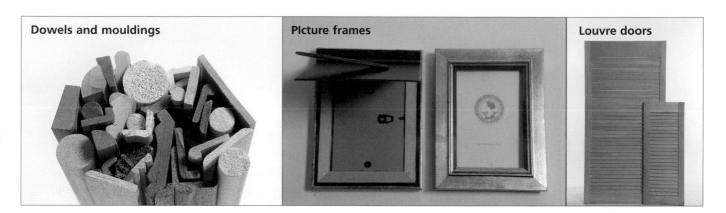

Dowels and mouldings

Picture frames

Louvre doors

Quotas and Export Restrictions

Where there are concerns about a species in trade, export quotas are established to limit trade volume. These are set by either individual Parties unilaterally or by the Conference of the Parties, and are generally in place from 1 January to 31 December. Quotas can be subject to annual change in relation to updated information regarding the status of the species concerned and trade levels.

Quotas for Ramin exports from Malaysia and Indonesia are currently in place. Updates of these quotas can be found at:

www.cites.org/eng/resources/quotas/index.shtml

Some countries may put in place other restrictions on export, for example, Indonesia limits harvesting of Ramin to concessions that have received a certification of Sustainable Forest Management. Furthermore, they restrict the number of companies that are authorised to export Ramin. The CITES Secretariat informs Parties of such restrictions by means of formal Notifications.

Notifications to the Parties can be found at:

www.cites.org/eng/notif/valid.shtml

Import Suspensions

Member States of the European Union and other countries may, at times, put in place import suspensions for certain species and/or regions.

Information on current European Union import suspensions can be found at:

http://ec.europa.eu/environment/cites/legislation_en.htm

It is important to note that the published import suspensions are reversible at any moment if new information is received.

Trade routes – Global

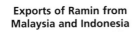

➡ **Exports from range States**

➡ **Re-exports**

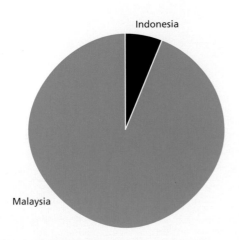

**Exports of Ramin from
Malaysia and Indonesia**

Indonesia

Malaysia

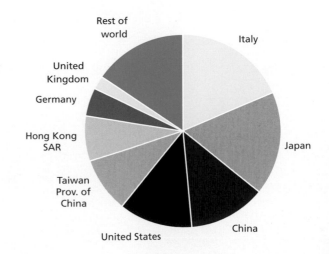

Ramin destinations

Rest of
world

United
Kingdom

Germany

Hong Kong
SAR

Taiwan
Prov. of
China

United States

Italy

Japan

China

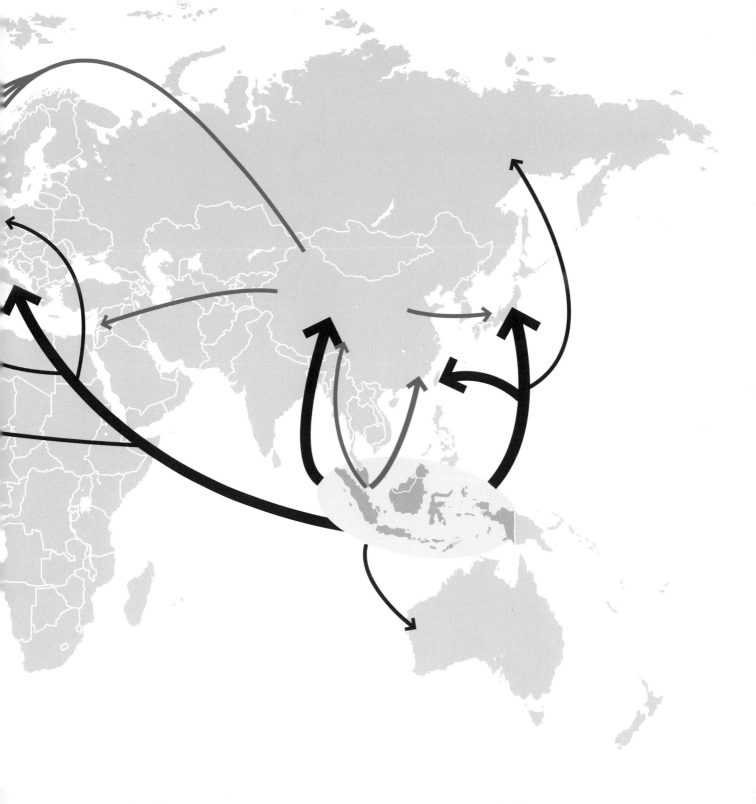

Countries re-exporting Ramin

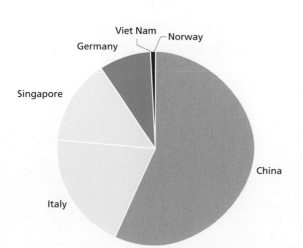

Viet Nam
Norway
Germany
Singapore
Italy
China

Ramin products re-exported

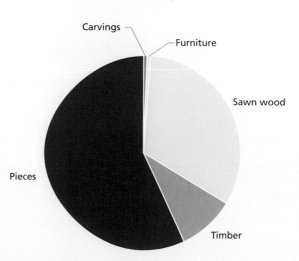

Carvings
Furniture
Sawn wood
Pieces
Timber

Trade routes – Europe

➡ **Exports from range States**
➡ **Re-exports**

Thirty percent of exports from range States are exported to Europe, of which 92% are imported from Malaysia and 8% from Indonesia

Re-exports from Europe account for 22% of global re-exports

USA, Canada

European destinations

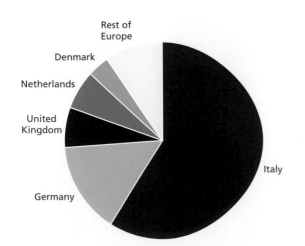

Rest of Europe

Denmark

Netherlands

United Kingdom

Germany

Italy

Export products into Europe

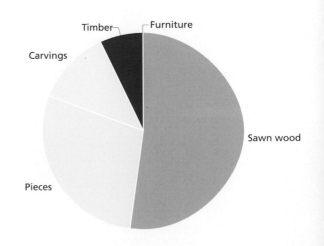

Timber

Carvings

Furniture

Sawn wood

Pieces

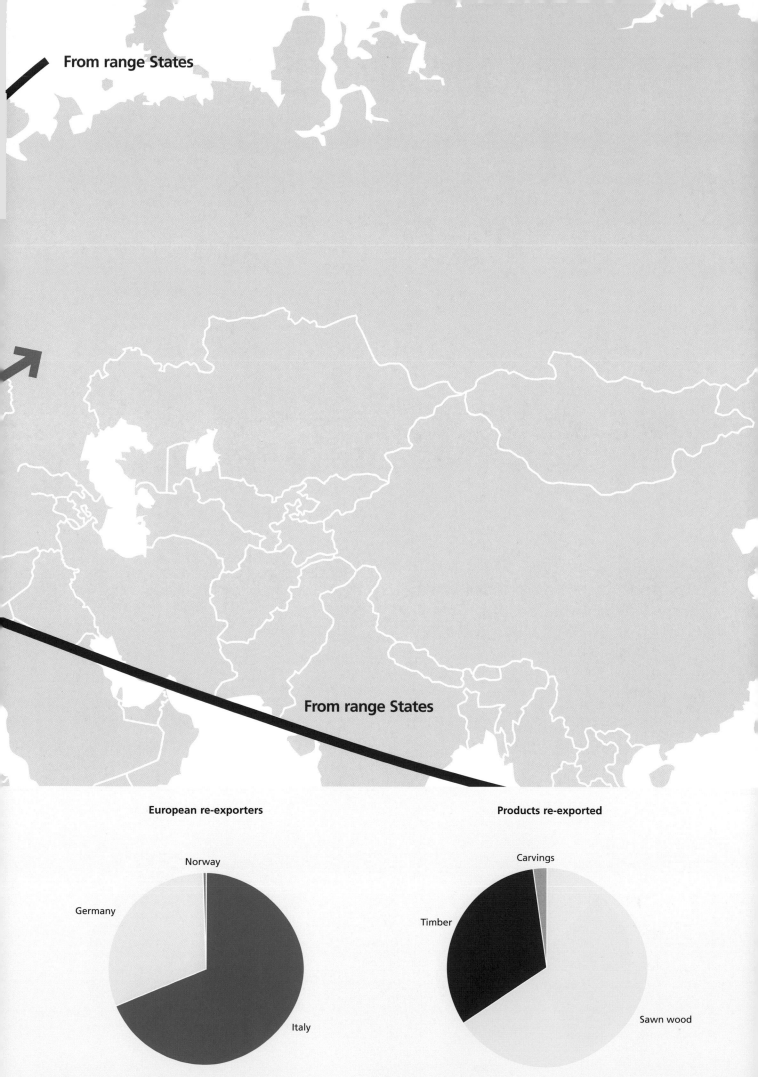

From range States

From range States

European re-exporters

Norway

Germany

Italy

Products re-exported

Carvings

Timber

Sawn wood

Trade routes – Asia

→ **Exports from range States**

→ **Re-exports**

Europe

South Africa

Fifty one percent of exports from range States are exported within Asia, of which 98% are imported from Malaysia and 2% from Indonesia

Asian re-exports account for 78% of global re-exports

Asian destinations

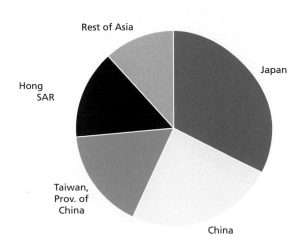

Rest of Asia

Japan

Hong SAR

China

Taiwan, Prov. of China

China

Export products into Asia

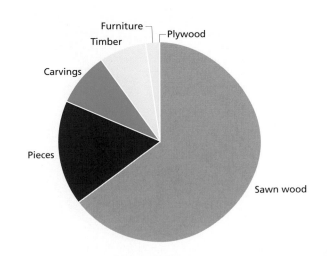

Furniture

Timber

Plywood

Carvings

Pieces

Sawn wood

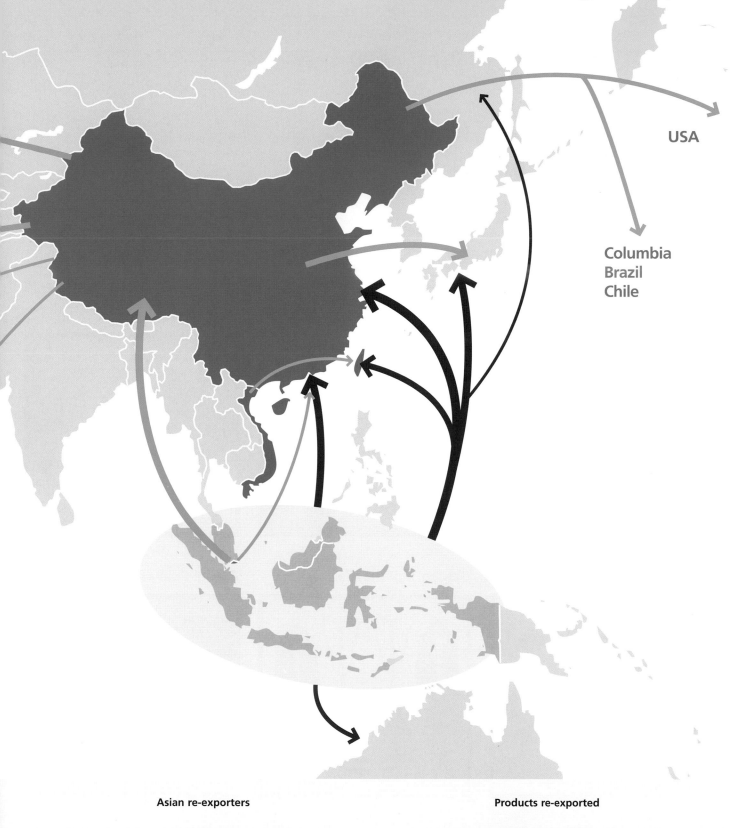

USA

Columbia
Brazil
Chile

Asian re-exporters

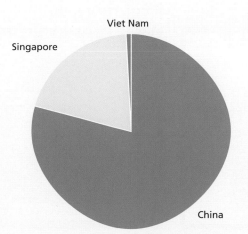

Viet Nam

Singapore

China

Products re-exported

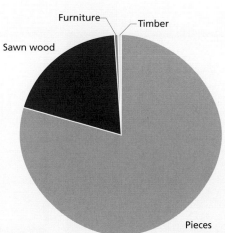

Furniture — Timber

Sawn wood

Pieces

Parts and derivatives regulated

The following products, which may be made of Ramin, are seen in international trade. Those products regulated by CITES are highlighted in **red**.

Due to the continual updating of specific codes, tariff **chapter codes** are listed rather than full individual tariff codes. A more detailed guide to tariffs codes can be found in your current edition of International Integrated Tariffs. The photographs used to illustrate the products are not, in all cases, Ramin. There is little or no trade in plywood, veneer or parquet flooring.

Logs or rough wood

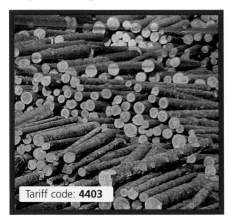

Tariff code: **4403**

Sawn wood >6mm thickness

Tariff code: **4407**

Veneer sheets <6mm thickness

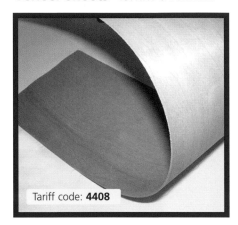

Tariff code: **4408**

Clothes hangers

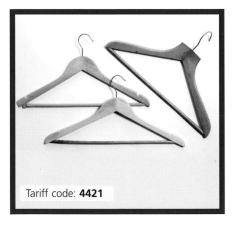

Tariff code: **4421**

Garden Tool Handles

Tariff code: **4417**

Picture Frames

Tariff code: **4414**

Plywood

Tariff code: **4412**

Louvre doors

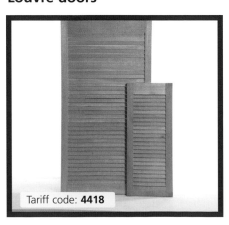

Tariff code: **4418**

Parquet flooring

Tariff code: **4409** and **4418**

Mouldings – Furniture pieces

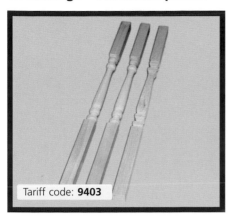

Tariff code: **9403**

Mouldings – Picture frames

Tariff code: **4414**

Snooker cues

Tariff code: **9504**

Mouldings

Tariff code: **4415**

Mouldings – Dowels

Tariff code: **4415**

Furniture

Tariff code: **9503**

Table and kitchenware

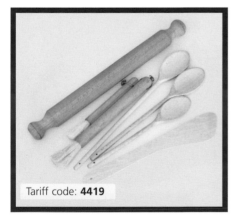

Tariff code: **4419**

Venetian blinds

Tariff code: **4421**

Door frames

Tariff code: **4418**

Other: Dolls house furniture

Tariff code: **9503**

Other: Paintbrushes

Tariff code: **9603**

Other: Light pull ends

Tariff code: **4421**

Identification

Wood can be identified using one or more of the following techniques:

- **Morphological identification** – where the characteristics (type, distribution and arrangement) of the wood's vessels (pores) are used to identify the wood sample. This can use either:

  - **Macroscopic characteristics** – visible to the unaided eye or with the help of 10x power or above hand lens. This can be carried out in the field; OR

  - **Microscopic characteristics** – these are too small to be seen with the unaided eye or with a hand lens. A microscope is required. This is carried out in a laboratory.

- **Chemical identification** – this includes testing for specific chemicals in a sample (chromatography) or looking at the sample's DNA sequence to identify the species (DNA analysis). These tests have to be carried out in a laboratory by an expert.

When identifying a wood sample it is necessary to understand which plane (referred to as "face", "section" or "surface") you are looking at. The transverse or cross section is the most useful for viewing the wood's vessels (below):

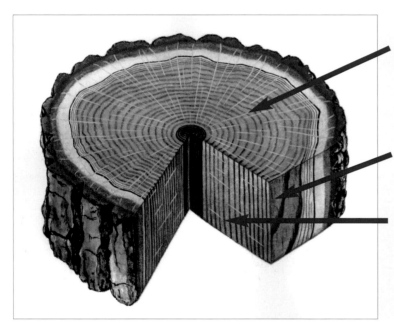

Transverse plane
This plane is sometimes also called a cross-sectional plane or simply a cross section, end or cross grain and often provides the most useful information about the distribution, type and arrangement of the wood vessels.

Tangential plane
A longitudinal plane at right angles to the radius of the stem.

Radial plane
A longitudinal plane along the radius of the stem.

Morphological identification

Macroscopic identification

Equipment

The following equipment is needed when identifying a wood sample:

- Hand lens (10x or above magnification)

- Sharp hand saw, razor blades, pen knife or other cutting equipment – to take samples and prepare the sample for identification

- Sandpaper (fine to medium coarseness) – to prepare the wood surface to view the macroscopic characteristics

- Camera or mobile phone - to take pictures of the samples or shipment

If you are examining a large shipment you may also need:

- Face mask, goggles & gloves – to protect against wood dust and splinters

- Tape measure and weighing equipment – to check the sample's weight and dimensions against any paperwork or permits

- Plastic bags – to collect samples for identification back in the laboratory or to send to experts

Tools to help in identification:

- There are a now a number of dedicated CITES manuals and computer-based identification tools to help identify your wood sample. These are identified in the resource section of this guide. We have included a copy of one CD–ROM based guide (CITES*wood*ID) with this manual and will introduce you to its use.

CITES I-II-III Timber Species Manual

How to prepare a wood sample for identification

1. If identifying larger pieces of sawn timber or logs it may be easier to cut smaller samples from them (at least 2-3cm^3). This will make the sample easier to handle, enable you to find the transverse plane and move it, if necessary, to a location with good lighting (i.e. natural lighting, not artificial lighting) and take the sample to your office to check against a paper or computer identification guide if required.

2. Ensure the wood surface is freshly sliced or sanded down and does not have rough or jagged edges as this makes it hard to see the wood's vessels.

3. Hold the sample firmly in one hand ensuring your hand/fingers are below the section being sliced.

4. Make a clean slice along the surface of the sample. Do this by holding the razor blade with your fingers and pushing it with your thumb. Ensure you slice away from your body. Slice deep enough to reveal the wood's vessels.

5. Put the hand lens up to your eye. Move the sample back & forward keeping the hand lens next to your eye until you can see some pores or vessels in focus. If you cannot see any pores or vessels you may not be looking at the characteristics on the transverse plane (the different planes are shown in the figure on page 14), you may not have sliced a deep and long enough section to see the vessels clearly or you may have the hand lens the wrong way round to your eye!

6. Re-slice the sample throughout the identification process as a fresh cut exposes the vessels more clearly.

7. Once you have done all of the above you are ready to check your sample against one of the identification tools previously mentioned above.

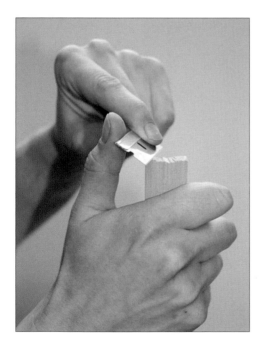 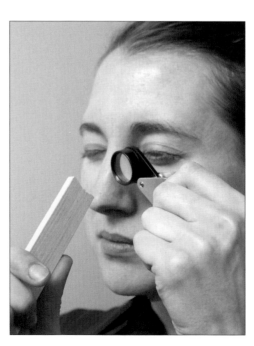

Identification guides

One of the most useful identification guides available is the computer-based morphological key, CITES*wood*ID Version 2008–2 and its updates (Richter, Gembruch and Koch, 2008). The guide uses a database of photographs of macroscopic characteristics with descriptions and notes to aid identification. Working by a process of elimination, the guide eliminates these timbers one by one until left with a remaining name.

The database contains:

- The CITES-listed timbers utilized for their wood products. CITES-listed plants/trees utilized for non-wood products have not been included.

Bulnesia sarmientoi	*Bulnesia sarmientoi*	CITES III *
Caesalpinia echinata	Brazil wood	CITES II
Cedrela odorata	Cedro	CITES III
Dalbergia nigra	Brazilian rosewood	CITES I
Dalbergia retusa	Cocobolo	CITES III
Dalbergia stevensonii	Honduras rosewood	CITES III
Fitzroya cupressoides	Alerce	CITES I
Gonystylus spp.	**Ramin**	**CITES II**
Guaiacum spp.	Lignum Vitae	CITES II
Pericopsis elata	Aformosia	CITES II
Platymiscium spp.	Granadillo	CITES II
Swietenia spp.	Mahogany	CITES II

* **Note:** This species is subject to a listing proposal at the forthcoming meeting of the CITES Conference of the Parties and may be up-listed to CITES Appendix II.

- 45 traded timbers which can be easily mistaken for CITES protected timbers due to a very similar appearance and/or structural pattern.

How to use CITES*wood*ID to identify Ramin

The terminology used in this programme may appear a little daunting at first if you have little or no knowledge of wood anatomy terminology. However, practice just using the visuals until you come to understand the terms being used.

- To start **CITESwoodID** insert the CD-ROM into your computer.

- The programme may:

  – Start up automatically – choose a language then click OK. A title page will appear. Click anywhere on this page to start the programme.

  – Not start up automatically – click on disc symbol, then double click on the red triangle and choose the appropriate language and click OK. Click anywhere on the title page to start the programme.

Intkey.exe

- The next screen is divided into four windows (below):

  – Window 1a = a list of characteristics relating to wood anatomy

  – Window 1b = used characteristics

  – Window 2a = remaining taxa i.e. the names of the CITES and non-CITES-listed timbers in the database

  – Window 2b = eliminated taxa

At the start only Windows 1b and 2b should be empty.

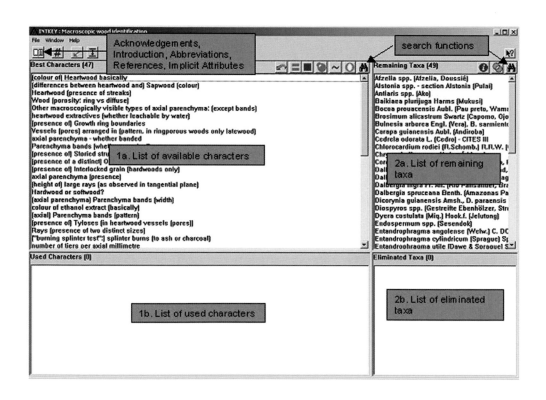

- Click on the symbol ▬ 'Best order' to rank the characteristics in Window 1a in the most useful order.

- Click once on the first characteristic in Window 1a, in this case 'heartwood (colour of) basically'. A new window appears presenting various characteristics to choose from. Highlight whichever characteristic you think most appropriate to your sample. In this case the colour of your sample is yellow so click and highlight '3. yellow' and then click on 'OK'.

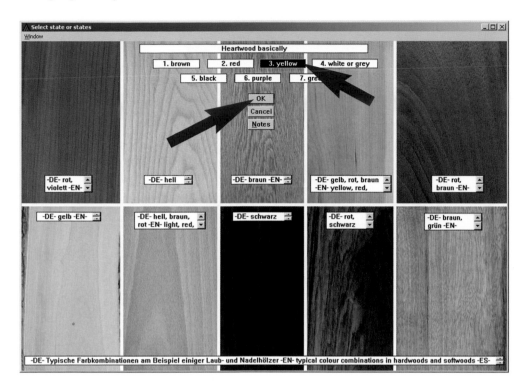

- As you work your way down the list of characteristics in Window 1a you will see that Window 1b and Window 2b fill up with the characteristics, and thus taxa names, that you have eliminated from the process. Window 2a lists the remaining taxa that your sample could be.

- Continue to work your way through the characteristics in Window 1a.

  Note: Some may not be relevant to your sample or you may not have the time or equipment to verify the characteristics e.g. 'heartwood extractives (whether leachable by water)'

- At any time push the ▭▤ button (main menu) to access several text files (html format), i.e., an 'Introduction' to the database, 'Implicit attributes', 'Acknowledgements', 'References', 'Abbreviations' (internal codes used in the database), and 'Contacts and disclaimer'.

  Note: Independent of the program, the system provides access to several files in the directory EN\RTF such as:

    descr.rtf (document containing the descriptions of all taxa);

    chars-lg.rtf (character list with explanatory notes);

    chars-sh.rtf (character list without explanatory notes);

    names.rtf (list of the botanical and trade names of the timbers in the database).

The directory EN\WWW contains the character list with (chars.htm) and without (charsonly.htm) explanatory notes in HTML format.

- As you eliminate characteristics you will eventually be left with a small list of three or four taxa in Window 2a. At this point you can always activate the button or info button **ⓘ** which will allow you to look over a list of differences between the selected timbers and/or pull up images of the timbers onscreen for direct comparison.

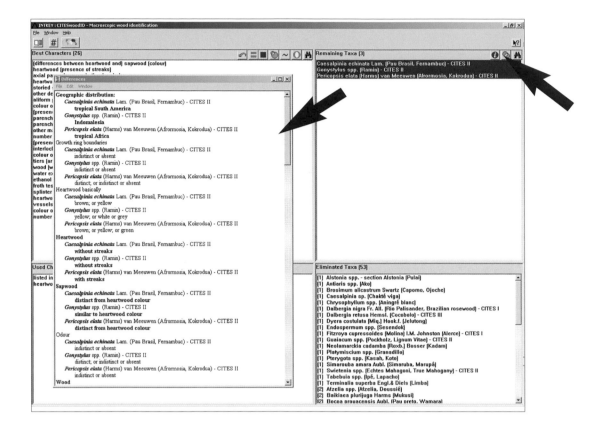

- Otherwise keep going until left with one name – *Gonystylus* spp. – **Ramin**. Again you can look at an image of this taxa by clicking on the info button **ⓘ** and verify against a picture of the transverse section showing ramin's distinctive winged vessels (circled below).

- At any time you can start the process from the beginning by clicking on the button

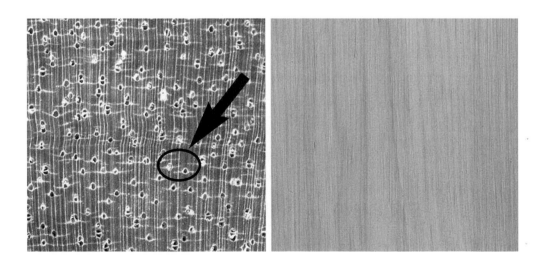

Differentiating between Ramin and similar looking woods

At any point during the identification process you can pull up pictures of the timbers you are assessing and check them against other species in the 'REMAINING TAXA' Window 2a. Either:

– Click on the double circle green button ⊘⊘ above Window 2a to display morphological differences between selected taxa. Double click on taxa to display images showing differences in characteristics.

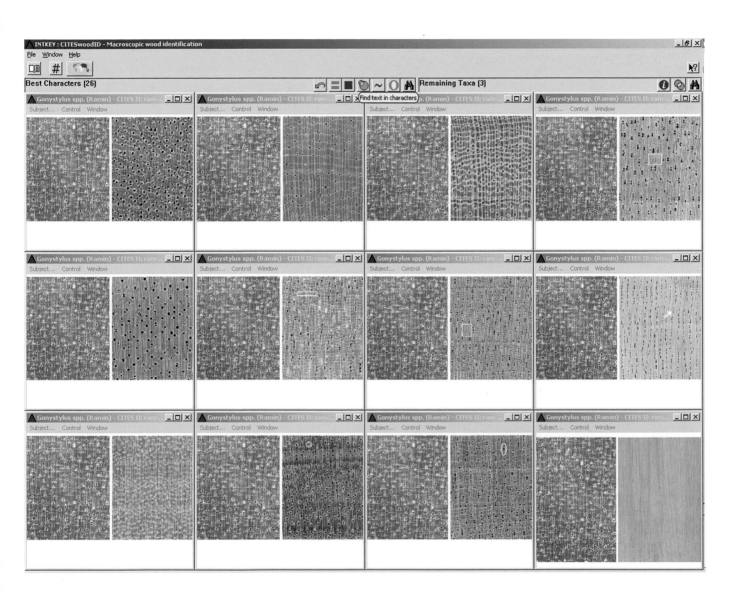

OR

– Click on the 'info' button 🛈 to pull up the 'taxon information' box. Here you can select info on full description, names, additional distinctive attributes and a diagnostic description. You can also click to get a picture up of the species in question or pictures comparing it to other similar looking woods in the database.

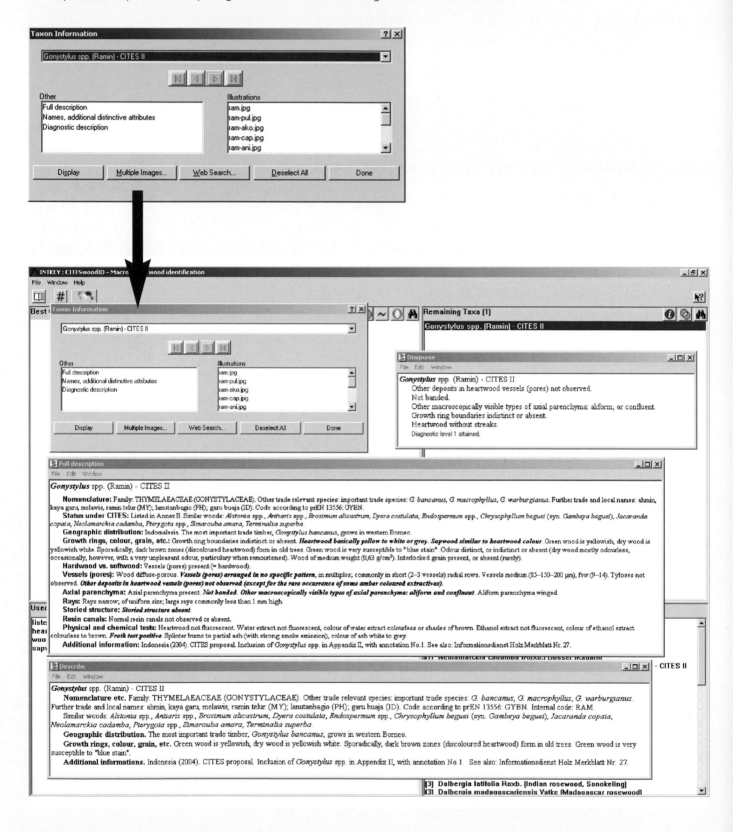

Microscopic identification

Using microscopic characteristics to identify timbers in trade is much more reliable, but is not practical in the field.

To carry out this test very fine slices of wood are taken on the transverse, radial and tangential planes. These slices are then set on a microscopic slide and examined under an optical microscope to observe the anatomical features. As characteristics are examined on all three planes and the wood is compared against existing vouchered (verified) samples, this test is a more accurate way of determining the identity of the wood.

 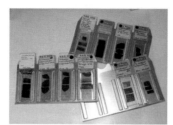  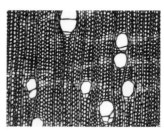

Chemical identification

Chemical identification is conducted in the laboratory as specialist equipment and expertise is required. The chemical tests utilised to identifying wood include:

- **Chromatography** – this is a laboratory technique that separates the chemicals within a sample and then measures the relative amounts of these chemicals to produce a profile for the species.

- **DNA analysis** – technique used to analyze genes and DNA.

There are 3 stages involved in the use of genetic techniques to identify CITES listed timbers:

1. Sample collection – performed at the point of inspection and can be carried out by the investigating officer with minimal training and equipment. Samples are then transferred to a laboratory.

2. DNA is extracted from the sample.

3. Genetic analysis. This is based on an industry standard technique and provides rapid unequivocal identification of Ramin DNA. The total laboratory stage takes approximately six hours from start to finish with the capacity to process 96 (or on some instruments 384) samples simultaneously.

Contacts:

UK → Trace Wildlife Forensics Network   email: info@tracenetwork.org
     → Royal Botanic Gardens, Kew   email: CAPS@kew.org

CITES documentation

The basic requirements for export of CITES Appendix II listed taxa is that a valid export permit should be issued by a Management Authority, following the advice of the Scientific Authority. Some countries, such as the member States of the European Union, apply stricter domestic legislation and require import permits in addition to the export permit.

Information on CITES requirements can be found at:
http://www.cites.org/eng/disc/how/shtml

Information on European Union CITES legislation can be found at:
http://ec.europa.eu/environment/cites/legislation_en.htm

Where an export and import document is required you should check for the following:

1. That the importer and exporter details match.

2. That the country of import and the country of export match.

3. That the issuing Management Authority of export is the same as shown in box 24 of the Import permit.

4. That the descriptions of the specimens are the same.

5. That the scientific name is the same on both documents.

6. That the common name is the same on both documents. Note: These may vary as some species have more than one common name.

7. If box 16 of the import permit is completed the export permit referred to **MUST** be the one presented.

8. That the CITES Appendix recorded on both documents are the same.

9. The source code may differ you will need to clarify this with your CITES Management Authority.

10. Although the quantity may vary, the import permit must cover (i.e. be equal to or less than) the amount recorded on the export permit.

CITES Standard permit/certificate form

Standard permit/certificate form

CITES CONVENTION ON INTERNATIONAL TRADE IN ENDANGERED SPECIES OF WILD FAUNA AND FLORA	**PERMIT/CERTIFICATE No.** ☐ EXPORT ☐ RE-EXPORT ☐ IMPORT ☐ OTHER:	Original	
		2. Valid until	

3. Importer (name and address)	4. Exporter/re-exporter (name, address and country)
3a. Country of import	_____ Signature of the applicant

5. Special conditions *For live animals, this permit or certificate is only valid if the transport conditions conform to the CITES Guidelines for transport or, in the case of air transport, to the IATA Live Animals Regulations*	6. Name, address, national seal/stamp and country of Management Authority

5a. Purpose of the transaction (see reverse)	5b. Security stamp no.

	7./8. Scientific name (genus and species) and common name of animal or plant	9. Description of specimens, including identifying marks or numbers (age/sex if live)	10. Appendix no. and source (see reverse)	11. Quantity (including unit)	11a. Total exported/Quota
A	7./8.	9.	10.	11.	11a.
	12. Country of origin * Permit no. Date		12a. Country of last re-export Certificate no. Date		12b. No. of the operation ** or date of acquisition ***
B	7./8.	9.	10.	11.	11a.
	12. Country of origin * Permit no. Date		12a. Country of last re-export Certificate no. Date		12b. No. of the operation ** or date of acquisition ***
C	7./8.	9.	10.	11.	11a.
	12. Country of origin * Permit no. Date		12a. Country of last re-export Certificate no. Date		12b. No. of the operation ** or date of acquisition ***
D	7./8.	9.	10.	11.	11a.
	12. Country of origin * Permit no. Date		12a. Country of last re-export Certificate no. Date		12b. No. of the operation ** or date of acquisition ***

* Country in which the specimens were taken from the wild, bred in captivity or artificially propagated (only in case of re-export)
** Only for specimens of Appendix-I species bred in captivity or artificially propagated for commercial purposes
*** For pre-Convention specimens

13. This permit/certificate is issued by:

_____ Place	_____ Date	_____ Security stamp, signature and official seal

14. Export endorsement: 15. Bill of Lading/Air waybill number:

Block	Quantity				
A					
B					
C		_____	_____	_____	_____
D		Port of export	Date	Signature	Official stamp and title

CITES PERMIT/CERTIFICATE No.

(as amended at CoP14)

CITES Standard permit/certificate form

Instructions and explanations:
These correspond to black numbers on the permit/certificate form

2. For export permits and re-export certificates, the date of expiry of the document may not be more than 6 months after the date of issuance (1 year for import permits).

3. **3a) MUST** be written in full.

4. The absence of the signature of the applicant renders the permit or certificate invalid.

5. Special conditions i.e. national legislations or conditions placed on the shipment by the issuing Management Authority.

   5a) Note: *The full list of abbreviations can be found on the reverse of the permit.*

   5b) Number of stamp affixed in Box 13.

6. The name, address and country of the issuing Management Authority should already be printed on the form.

7. e.g. *Gonystylus bancanus*.

8. e.g. Ramin.

9. Description of the specimen in trade, including any marks. e.g. louvre doors.

10. CITES Appendix listing and source e.g. II, W (Appendix II, Wild).

11. Units should conform to the most recent version of the *Guidelines for the preparation and submission of CITES annual reports*. For timber this should be m^3 or kg.

    11a) The total number of specimens exported in the current calendar year (1 Jan – 31 Dec) (including those covered by the present permit) and the current annual quota for the species (e.g. 500/1000). For timber this should be m^3 or kg.

12. The country in which the specimens were taken from the wild or artificially propagated.

    12a) Only to be completed in case of re-export of specimens previously re-exported. Country from which the specimens were re-exported before entering the country in which the present document is issued. Enter the number of the last re-export certificate and its date of issuance. If all or part of the information is not known, this should be justified in Box 5.

    12b) "Date of acquisition" only required for pre-Convention specimens.

13. To be completed by the official who issues the permit. Name must be written in full. Security stamp must be affixed in this block and must be cancelled by the signature of the issuing official and a stamp or seal.

14. To be completed by the official who inspects the shipment at the time of export or re-export. Enter the quantities of specimens actually exported or re-exported. Strike out the unused blocks.

15. Bill of Lading/ Air way-bill number.

European Community
Standard permit/certificate form

4	**1** Exporter/Re-exporter

PERMIT/CERTIFICATE **No**

- ☐ IMPORT
- ☐ EXPORT
- ☐ RE-EXPORT
- ☐ TRAVELLING EXHIBITION
- ☐ CITES PET OWNERSHIP

2. Last day of validity:

CITES Convention on International Trade in Endangered Species of Wild Fauna and Flora

COPY for the issuing authority

3. Importer

4. Country of (re)-export

5. Country of import

6. Authorized location for live wild-taken specimens of Annex A species

7. Issuing Management Authority

4

8. Description of specimens (incl. marks, sex/date of birth for live animals)	9. Net mass (kg)	10. Quantity

11. CITES Appendix	12. EC Annex	13. Source	14. Purpose

15. Country of origin

16. Permit No	17. Date of issue

18. Country of last re-export

19. Certificate No	20. Date of issue

21. Scientific name of species

22. Common name of species

23. Special conditions

This permit/certificate is only valid if live animals are transported in compliance with the CITES Guidelines for the Transport and Preparation for Shipment of Live Wild Animals or, in the case of air transport, the Live Animals Regulations published by the International Air Transport Association (IATA)

24. The (re-)export documentation from the country of (re-)export
- ☐ has been surrendered to the issuing authority
- ☐ has to be surrendered to the border customs office of introduction

25. The ☐ importation ☐ exportation ☐ re-exportation of the goods described above is hereby permitted.

Signature and official stamp:

Name of issuing official:

26. Bill of Lading / Air Waybill Number:

Place and date of issue:

27. For customs purposes only

Signature and official stamp:

Customs document
Type:

Quantity / net mass (kg) actually imported or exported	Number of animals dead on arrival

Number:

Date:

European Community
Standard permit/certificate form

Instructions and explanations:
These correspond to black numbers on the permit/certificate form

1. Name & address of person, persons, or company exporting the shipment.

2. Permit expiry date.

3. Name & address of person, persons, or company receiving the shipment.

4. country arriving from e.g. Malaysia.

5. Country entering into e.g. UK.

6. Location for live wild-taken specimens of Annex A species.

7. Address of Management Authority that issued the permit.

8. Whether the specimen is a live plant, a part or derivative – including any specific labels or markings.

9. Net mass (kg). For timber this should be m³ or kg.

10. Quantity.

11. CITES Appendix.

12. EC Annex Listing.

13. The source of the specimen, e.g.: W (wild) – this specimen comes from a wild non-regulated environment.

    The full list of abbreviations to be used can be found on the reverse of the Permit.

14. What the specimen(s) is/are to be used for.

    For example: T (commercial) – may be sold for commercial purposes,

    P (personal use) – only for own use.

    The full list of abbreviations can be found on the reverse of the Permit.

15. Where the specimen was removed from its natural environment.

16. If completed on the import permit, the export permit referred to in this box **MUST** be the permit presented.

17. Date of issue of permit in Box 16.

18. Country of last re-export if different to Box 15. That is the country from which the specimens were re-exported before entering the country in which the present document was issued. Including the:

19. Permit number for the movement, and,

20. The Permit date of issue.

21. Latin name, e.g. *Gonystylus bancanus.*

22. e.g. Ramin.

23. Any special conditions imposed by exporting country.

24. (Re)-export documentation from country of (re)export.

    Note: Not all EU Member States complete box 24.

25. The stamp and signature of the issuing officer and the date of issue of the permit.

    Note: The signature should cancel through the CITES security stamp if one is used.

26. Bill of lading/ Air way-bill number.

27. Export/ re-export/ import endorsement **MUST** be completed by the officer inspecting the documents on the export/ import/ re-export.

Timber measurement

Timber is traded in many different forms and often the amounts recorded on invoices, permits, etc are in different unit codes (e.g. carvings, sawn wood, etc). To assist you when verifying the quantity invoiced on the shipment's documents matches the quantity recorded on the CITES permit or certificate we recommend you use the following formulas.

A number of the units used are specific to certain countries and may not be used in your country e.g. board feet is a unit of volume often used in the USA and Canada.

For assistance, ask the trader whether they use any standard conversion rates or contact your CITES Scientific Authority or your local or national forestry/plant health agency. These conversions should also be made by the importer or exporter so that the total quantity of CITES regulated material recorded on the shipping documents is expressed in the same unit of measurement found on the CITES documentation.

Note: Conversion rates are taken from USDA CITES I-II-III TImber Species Manual (2006).

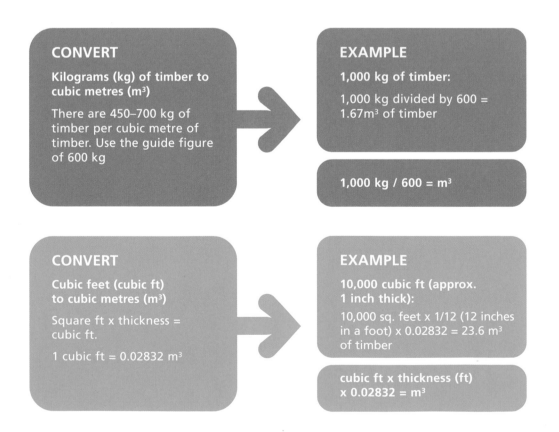

CONVERT

Kilograms (kg) of timber to cubic metres (m³)

There are 450–700 kg of timber per cubic metre of timber. Use the guide figure of 600 kg

EXAMPLE

1,000 kg of timber:

1,000 kg divided by 600 = 1.67m³ of timber

1,000 kg / 600 = m³

CONVERT

Cubic feet (cubic ft) to cubic metres (m³)

Square ft x thickness = cubic ft.

1 cubic ft = 0.02832 m³

EXAMPLE

10,000 cubic ft (approx. 1 inch thick):

10,000 sq. feet x 1/12 (12 inches in a foot) x 0.02832 = 23.6 m³ of timber

cubic ft x thickness (ft) x 0.02832 = m³

CONVERT

Square feet to square metres (m²)

Convert sq. ft to sq. metres (m²)

[sq. ft = length (ft) x width (ft)]

EXAMPLE

25,000 sq. feet of timber:

25,000 sq. ft x 0.0929 = 2,322.5 m² of timber

1 sq. ft x 0.0929 = m²

CONVERT

Square metres (m²) to cubic metres (m³)

Convert sq. metres to cubic metres (m³)

EXAMPLE

25,000 m² of timber (veneer) (0.6mm thick):

25,000 m² x 0.0006m = 15 m³ of timber

m² x thickness = m³

CONVERT

Volume of a cylinder (inches) to cubic metres (m³)

Convert volume of cylinder inches to cubic metres.

N.B. π (3.14) x (radius in inches)² x (length in inches) x (total number of dowels) = cubic inches.

EXAMPLE

100,000 dowels (1/4 inches diameter) x 16 inches in length:

Radius = 1/2 diameter ➔ 1/4=0.25 x 1/2 = 0.125 (3.14) x (0.125)² x 16 inches x 100,000 = 78,500 cubic inches.

(78,500 cubic inches) x 0.0000164 = 1.287 m³

(cubic inches) x (0.0000164) = m³ of dowel

CONVERT

Board feet (usually expressed as pie tablares (PT)) to cubic metres (m³)

There are 424 PT per cubic metre

EXAMPLE

1,000 board feet (PT) of timber:

1,000 PT of timber divided by 424 = 2.36 m³ of timber

1,000 PT / 424 = m³

Key resources

Identification

CITESwoodID: Descriptions, illustrations, identification, and information retrieval. In English, German, French, and Spanish. Version: 2008-2.
H.G. Richter, K. Gembruch, G. Koch (2005 onwards)
Federal Research Institute for Rural Areas, Forestry and Fisheries (vTI), Hamburg, Germany.
Enables the user to identify by means of macroscopic characters traded timbers which are regulated by CITES.

United States Department of Agriculture
CITES I-II-III Timber Species Manual
Provides the procedures for the enforcement of CITES for Appendix I, II and III tree species.
www.aphis.usda.gov/import_export/plants/ manuals/ports/downloads/cites.pdf

CITES Identification Guide – Tropical Woods (2002)
Guide to the identification of tropical woods controlled by CITES. Wildlife Enforcement and Intelligence Division, Enforcement Branch, Environment Canada.
www.cws-scf.ec.gc.ca/enforce/pdf/Wood/ CITES_Wood_Guide.pdf

Commercial timbers: descriptions, illustrations, identification, and information retrieval
H. G. Richter and M. J. Dallwitz (2000 onwards)
www.biologie.uni-hamburg.de/b-online/wood/english/index.htm

Trace Wildlife Forensics Network   **www.tracenetwork.org**

CITES Information

CITES Secretariat website   **www.cites.org**
For updates on quotas, restrictions, training and identification materials, courses and workshops.

European Commission CITES website   **http://ec.europa.eu/environment/cites/home_en.htm**

UK CITES Management Authority   **www.defra.gov.uk/animalhealth/cites/**

UK Scientific Authority for Plants   **www.kew.org/conservation/cites-ind.html**

UK Border Agency – CITES Team   **Lap.cites@hmrc.gsi.gov.uk**

U.S. Fish & Wildlife Service Management & Scientific Authorities website   **www.fws.gov/international/DMA_DSA/ CITES/CITES_ home.html**